Robins

by Martha London

FOCUS READERS

PIONEER

www.focusreaders.com

Focus Readers is distributed by North Star Editions:
sales@northstareditions.com | 888-417-0195

Produced for Focus Readers by Red Line Editorial.

Photographs ©: Shutterstock Images, cover, 1, 7, 8, 17, 21 (nest); iStockphoto, 4, 11, 14, 18; Leonard Lee Rue III/Science Source, 13; Red Line Editorial, 21 (chart)

Library of Congress Cataloging-in-Publication Data
Names: London, Martha, author.
Title: Robins / by Martha London.
Description: Lake Elmo, MN : Focus Readers, [2021] | Series: Neighborhood safari | Includes index. | Audience: Grades 2-3
Identifiers: LCCN 2020002235 (print) | LCCN 2020002236 (ebook) | ISBN 9781644933558 (hardcover) | ISBN 9781644934319 (paperback) | ISBN 9781644935835 (pdf) | ISBN 9781644935071 (ebook)
Subjects: LCSH: Robins--Juvenile literature.
Classification: LCC QL696.P288 L66 2021 (print) | LCC QL696.P288 (ebook) | DDC 598.8/42--dc23
LC record available at https://lccn.loc.gov/2020002235
LC ebook record available at https://lccn.loc.gov/2020002236

Printed in the United States of America
Mankato, MN
082020

About the Author

Martha London writes books for young readers. When she's not writing, you can find her hiking in the woods.

Table of Contents

Chapter 1

Looking for Worms

A robin hops across a lawn. She dips her head and catches a worm. Then she flies to her nest and feeds her babies.

Robins are songbirds. They often live near forests and fields. Trees give robins **shelter**. Fields have insects and worms for them to eat. Many robins live in cities, too. People can see them in parks or yards.

Chapter 2

Orange Chests

Robins have orange chests. Their backs and wings are gray or brown. Robins have white feathers under their tails.

Robins are approximately 10 inches (25 cm) long. They have hard beaks. They have thin legs. Their feet are brown. Each foot has four toes. Robins use their toes to grip branches.

Fun Fact

Robins can make several different sounds. They can **chirp** and **whistle**.

Building Nests

Robins build nests on trees, cliffs, or buildings. They make a new nest each time they lay eggs. Each nest is shaped like a cup. Robins use grass and sticks to make this shape. Then they add mud to the sides. The mud makes the nest **sturdy**.

Chapter 3

A Bird's View

Robins have excellent eyesight. A robin's eyes are on the sides of its head. As a result, the bird can see a wide area around itself.

Robins watch for **predators**. They also look for food. Robins can spot tiny movements on the ground. They use their beaks to pick up insects and worms. Robins also pick berries.

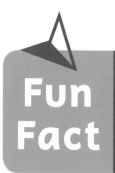

Fun Fact

Robins can fly up to 36 miles per hour (58 km/h).

Chapter 4

Flocks and Chicks

In fall and winter, robins form large **flocks**. They look for berries. Some robins **migrate**. They fly to places with more food.

Other robins stay in the same place all year. They fluff their feathers to stay warm. Robins leave their flocks in spring and summer. They find places to nest. When their eggs hatch, the parents care for the chicks.

Fun Fact

Robin eggs are bright blue.

Life Cycle

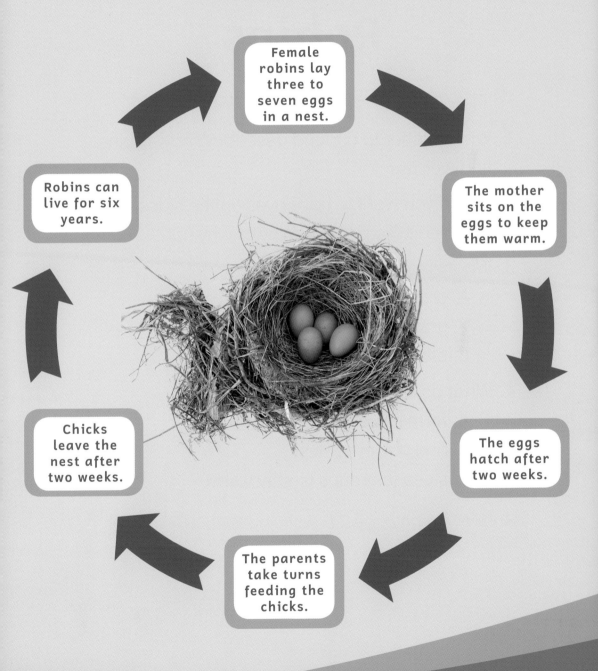

Female robins lay three to seven eggs in a nest.

The mother sits on the eggs to keep them warm.

The eggs hatch after two weeks.

The parents take turns feeding the chicks.

Chicks leave the nest after two weeks.

Robins can live for six years.

FOCUS ON
Robins

Write your answers on a separate piece of paper.

1. Write a letter to a friend describing how robins find worms.

2. Would you want to have eyes on the sides of your head like a robin? Why or why not?

3. What color is a robin's chest?
   - A. orange
   - B. gray
   - C. brown

4. What do robins do in the summer?
   - A. form large flocks
   - B. find places to nest
   - C. sleep in groups to stay warm

Answer key on page 24.

Glossary

chirp
To make a short, high sound.

flocks
Groups of birds that live together.

migrate
To move from one place to another when the seasons change.

predators
Animals that hunt other animals for food.

shelter
Protection from danger or weather.

sturdy
Not easily broken.

whistle
To make a high, airy sound.

To Learn More

BOOKS

Amstutz, Lisa J. *Robins*. North Mankato, MN: Capstone Publishing, 2016.

Christelow, Eileen. *Robins! How They Grow Up*. Boston: Houghton Mifflin Harcourt, 2017.

NOTE TO EDUCATORS

Visit **www.focusreaders.com** to find lesson plans, activities, links, and other resources related to this title.

Index

Answer Key: **1.** Answers will vary; **2.** Answers will vary; **3.** A; **4.** B